0712330100

OPTICAL VIDEO DISC TECHNOLOGY AND APPLICATIONS

recent developments in the USA

R Barrett

Library and Information Research Report 7

ABSTRACT

This report refers to a visit made by the author to the United States of America in August and September 1981. The objectives of the visit were: (1) to visit DiscoVision Associates to solve problems of formatting of material on 35-mm sprocketed roll film to make it compatible with requirements for video disc mastering; (2) to discuss with selected bodies progress in preparing material for video disc mastering and experience of use; and (3) to assess progress, if any, in adapting the analogue video disc for digital applications. The report indicates a disappointing situation, in that quality control problems are still restricting yield, even for analogue video discs, and very little consideration of digital applications has yet been made.

Library and Information Reseach Reports are published by the British Library and distributed by Publications Section, British Library Lending Division, Boston Spa, Wetherby, West Yorkshire LS23 7BQ.

621.3883 3
BAR

ISBN 0 7123 3010 0
ISSN 0263-1709

Typeset by The Palantype Organisation Limited, 4 North Mews, London WC1N 2JP and printed in Great Britain by DND Business Services, Tadcaster, North Yorkshire.

CONTENTS

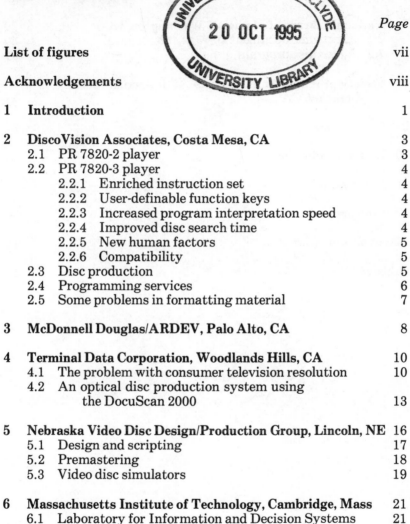

	Page
List of figures	vii
Acknowledgements	viii
1 Introduction	1
2 DiscoVision Associates, Costa Mesa, CA	3
2.1 PR 7820-2 player	3
2.2 PR 7820-3 player	4
2.2.1 Enriched instruction set	4
2.2.2 User-definable function keys	4
2.2.3 Increased program interpretation speed	4
2.2.4 Improved disc search time	4
2.2.5 New human factors	5
2.2.6 Compatibility	5
2.3 Disc production	5
2.4 Programming services	6
2.5 Some problems in formatting material	7
3 McDonnell Douglas/ARDEV, Palo Alto, CA	8
4 Terminal Data Corporation, Woodlands Hills, CA	10
4.1 The problem with consumer television resolution	10
4.2 An optical disc production system using the DocuScan 2000	13
5 Nebraska Video Disc Design/Production Group, Lincoln, NE	16
5.1 Design and scripting	17
5.2 Premastering	18
5.3 Video disc simulators	19
6 Massachusetts Institute of Technology, Cambridge, Mass	21
6.1 Laboratory for Information and Decision Systems	21
6.2 The Architecture Machine Group	24

7   National Library of Medicine, Washington, DC   29
    7.1   Electronic document storage and retrieval   30
    7.2   Video disc program   31

8   Society of American Archivists' 45th Annual Meeting,
        Berkeley, CA   33

9   Conclusions   37

10 Recommendations   38

11 References   39

12 Stop press   40

List of abbreviations   41

Other reports   42

LIST OF FIGURES

Page

1 Model VMT-2000: a video frame memory and
   high-resolution terminal (Terminal Data Corporation)   11

2 DocuScan: a high-speed document digitizer
   (Terminal Data Corporation)   14

ACKNOWLEDGEMENTS

The visits were made possible by a grant from the British Library Research and Development Department under its Study Visit Overseas scheme.

The data presented in the report and opinions expressed by the author are based on information freely given in the form of discussions and technical papers by staff of the following establishments, to whom due acknowledgement is made.

DiscoVision Associates

McDonnell Douglas Corporation

Massachusetts Institute of Technology

National Library of Medicine

Nebraska Video Disc Design/Production Group

Public Archives of Canada

Terminal Data Corporation

1. INTRODUCTION

In the report to the British Library Research and Development Department entitled *Developments in optical disc technology and the implications for information storage and retrieval*[1] the author discussed some possible non-consumer applications of the optical video disc which might be of interest in library and other information-handling systems. The interest shown by workers in these fields has been tremendous, as evidenced by the number of lectures and demonstrations requested of the author.

The report recommended that, since it seems fairly certain that both the optical video disc and the digital optical recording (DOR) disc will have a significant part to play in library and other information systems in the relatively near future, the British Library should be involved at an early stage, both in evaluation of the prospective systems and in research and development into what can and cannot be done. It is essential that the technology is understood in this way in order that meaningful applications studies can be carried out without being hampered by equipment problems, as has happened so often in other fields. Such a programme would be an obvious extension of the remote access development at Hatfield Polytechnic since an optical video disc player and at least some of the necessary computer facilities are available there.

During the final phase of the remote access project at Hatfield[2] preliminary consideration was given to the preparation of an optical video disc which, in conjunction with the player, could replace the microfiche storage and retrieval unit. The database was currently the Polytechnic library catalogue which was produced on computer output microform (COM) fiche and updated fortnightly. In the first instance it was decided to transfer the catalogue entries to disc in straight video mode and in order to ensure legibility the entries were to be restricted to two entries per video frame. The preferred format for disc mastering was 35-mm roll film which had to be sprocketed to be compatible with 525-line American standard tele-cine operation. COM bureaux normally only supply fiche but Eurocom Data has been very cooperative in experimenting with the required format. Rank Cintel was also very helpful in allowing the author to run test pieces of film through its tele-cine machine. However, it became clear that, in order to optimize the film format for disc mastering, a visit to DiscoVision Associates (DVA) in California was necessary,

1

since in the summer of 1981 it was the only company able to provide a disc production service.

A visit to the USA was therefore arranged for the above purpose and also to discuss progress in disc preparation and selected applications studies being carried out at a number of institutions. The visits are reported on in the following sections.

2. DISCOVISION ASSOCIATES, COSTA MESA, CA

Visits were made both to the headquarters and laboratories of DVA and to the video disc production plant. At the present time DVA's main concentration is undoubtedly on straightforward video disc production for the consumer market, but it is extending its interactive training involvement and has made further developments in its range of players.

2.1 PR 7820-2 player

The PR 7820-2 consists basically of internal modifications to the standard PR 7820, which may be either factory-installed at initial order time or retro-fitted to existing players at a DVA depot. It is also available with a Universal External Interface (UEI) communications adapter for linking the player to a digital computer with command and control of the video disc playback process. User software, resident in the attached host computer, can thus utilize DVA video disc equipment as an intelligent peripheral.

The UEI communications adapter provides either a standard RS232C serial interface or an IEEE 488 parallel interface for the PR 7820-2 and contains all hardware and software necessary for serial or parallel communication with a host computer, modems or a terminal. The UEI fully supports the PR 7820-2 player 8-bit parallel protocol, thus providing a complete path between a host computer and the player for send/receive/acknowledgement interfacing of data and status information. The UEI can receive and buffer a series of player commands from the host, relaying them to the player on command, and it continuously monitors either the RS232C or IEEE 488 and PR 7820-2 data paths as part of its regular operating procedure. Diagnostic information is sent to the remote host device as necessary.

The PR 7820-2 significantly increases frame search capability for users. Access time is on average 2½ times faster than the standard PR 7820. Control of the audio channels and screen display is provided on the PR 7820-2, and player operational status, as well as full random access memory contents, can be read by the attached computer.

2.2 PR 7820-3 player

DVA has also announced the PR 7820-3 player, which provides increased power for stand-alone applications. An enriched instruction set, user-definable function keys, six times faster program interpretation speed, improved disc search time and new human factors features expand the application capability over the standard PR 7820.

2.2.1 Enriched instruction set

With 22 new instructions, the PR 7820-3 can perform stand-alone applications previously possible only with an attached computer. The enhanced instruction set is particularly valuable for video disc users involved in exhibitions, point of purchase and training applications. For example, the new instructions make it possible for a continuous audio and video loop segment to be interrupted by a consumer at the point of purchase and provide direct access to specific product information at other locations on the video disc. Additionally, a time-out feature automatically returns the video disc system to continuous loop operations or to other pre-defined portions of the video disc after the consumer completes an inquiry. The PR 7820-3 also compares each input with pre-defined limits and identifies improper entries, allowing users to re-enter valid information.

2.2.2 User-definable function keys

The PR 7820-3 gives users the option of redefining the functions of 20 keys on the remote control unit or a user-supplied keyboard. Depending on the application, specific instructions can be custom-designed to broaden functions and provide greater ease of use. For example, consumer/student keyboards can be labelled with familiar terms like REPEAT, HELP, INDEX, START to simplify operation. Any special instruction can be used, provided it is pre-defined in the digital programs that perform the described function.

2.2.3 Increased program interpretation speed

Program interpretation speed for the PR 7820-3 is six times faster than that of the PR 7820-1. This feature is valuable in environments where the applications require multiple digit inputs (such as page or chapter numbers) with no visible delay to the user.

2.2.4 Improved disc search time

Average disc search time for the PR 7820-3 is 2½ times faster than for the PR 7820-1. This enhancement provides the user with more continuity when interacting with the video disc player.

2.2.5 New human factors

The increased programmability of the PR 7820-3 player makes practical many new features such as:

a) The capability to store the previous two locations from which the user has searched. This extremely valuable function enables the user to return automatically to material previously viewed without these locations being pre-defined in the digital program dump.

b) A valuable human factors feature which allows remote control unit entries to take effect at the PRESS of a key instead of the RELEASE. This is particularly useful when the video disc player is used in unattended public locations.

c) An optional no-charge feature automatically to initiate play when power is turned on or upon closing the lid if power is already on. This feature is useful in exhibit and point of purchase environments in which untrained employees can activate display units by simply turning on a power switch. Therefore, there should never be a reason to open the display unit after initial installation, except to change discs.

2.2.6 Compatibility

The PR 7820-3 is upwardly compatible with Model 1 and Model 2 programmed discs except in the following circumstances:

a) Because the Model 3 execution speed is faster, applications with pre-programmed timing loops based on the internal execution speed of the micro-processor should be examined to assess compatibility.

b) To avoid inadvertent or erroneous entries from a remote control unit when in program mode only REJECT and CLR/HLT will interrupt a running program. Therefore, programmed discs which have the user enter other commands to interrupt the program mode may not be compatible.

2.3 Disc production

DVA currently produces its discs by replication from a master in an injection moulding process. Preferred format for material provided by customers is sprocketed film, which is run through a Rank Cintel tele-cine machine to provide a 1-in video tape. Still material in particular is best presented on film to prevent jitter. Frame coding etc is added with frame numbering starting from the first frame presented, and a second video tape is made which is used to produce the master disc by laser ablation of a thin-film photoresist surface. Even if a properly formatted video tape is provided DVA still insists on the second stage and this can be a problem, especially in terms of increasing the error rate for digital recording.

In the production process a glass master disc is ground and polished and the photoresist is flowed on to an accurately controlled depth. This is then baked and put on the laser-cutting machine which is driven by the sub-master tape. A nickel-plated stamper is made, from which acrylic plastic copies are produced in the injection-moulding machines. An aluminium reflective layer is sputtered on in ovens and finally a protective plastic film is laid over the surface. Two discs are bonded together back-to-back to make a double-sided disc.

Even though they have now been producing discs for some time, DVA has still by no means solved the quality control problems and the yield is only about 80%. Each disc has to be played individually and inspected for dropouts. Dropouts are flaws in the disc's reflective surface such as tiny holes or scratches. A dropout could be the result of a flawed master, in which case all discs from that pressing would have the same defect. The manufacture of blanks and the stamping process itself are also to blame. With such a high reject rate the turn-round time at DVA, especially for special orders, has increased to 100 days.

2.4 Programming services

DVA offers a variety of services for producing video discs which include computerized editing, conversion of 35-mm or 16-mm film to video tape, scene-by-scene colour correction and image enhancement, high fidelity stereo sound sweetening and integration of digital control programs on to the program video tape.

DVA has also set up a section to deal with specialist activities such as that proposed by the author and shortly a special order facility is to be introduced to deal with small-run special requirements, in order to avoid the long turn-round time currently suffered.

At present any digital data must be presented in standard NTSC video format, with a capability, it is claimed, of accommodating about 170 bits/line. The data cannot be used to burn holes directly on to the disc because of the non-linearity of the channel. However, within a year DVA hopes to be able to accept digital data in industry standard magnetic tape format of 256 byte blocks and accommodate up to 300 Mbytes error-corrected per disc side.

DVA still has no real idea of the error performance characteristics of the discs and dropout errors are still causing considerable problems in frame coding — typically, it was claimed, one frame might drop in 15,000. This is why, for the first Video Patsearch discs, 54,000 frames could not be accommodated per side but 27,000 patent diagrams were put on to frames 1-27,000 then repeated on frames 27,001-54,000.

It has done some work on multiple-player synchronization and is currently developing a 15-unit synchronized controller for movie applications. This ought to be readily adaptable for straight multiple-disc frame retrieval in a document storage application.

2.5   Some problems in formatting material

A number of tests were carried out on the specimen filmed data provided by the author, which helped to clarify a number of problems. The tele-cine machine is set up according to the American SMPTE standard, which defines 'safe picture area' and takes account of the offset to accommodate a soundtrack. Also frames must be accurately located at four sprockets per frame, or frames will appear to drift out of area. DVA has done some work on type fonts and sizes for video disc applications and has produced a preliminary report[3] relating typewriter and typesetter font sizes to the safe picture area.

3. McDONNELL DOUGLAS/ARDEV, PALO ALTO, CA

As was previously reported[1], the Atlantic Richfield Corporation established a wholly-owned subsidiary, ARDEV Inc, to demonstrate the technical and commercial feasibility of a photographic-film-based video disc system over a wide range of applications. It saw the attractions of photographic technology over the more esoteric thin film approaches of other researchers as being the following:

a) The medium was based on known proven technology

b) There were low optical power requirements for recording

c) Mass or limited quantity replication was easy

d) Cost per copy would be very low

e) There was rapid turn-round from master to copy.

It pursued this line of research to the point of development of a system in which information was recorded on a silver-based film master disc by means of a lower power laser, using conventional processing and requiring only 45 minutes turn-round for a 30 minute playable master.

Replicated transmissive flexible discs (0.005 in thick) were produced by contact printing on to a non-silver-based medium and anhydrous ammonic development, giving linear grey-scale properties.

The discs were played back on a microprocessor-controlled disc stabilized player with random access and simultaneous or addressable multitrack read-out, using wide field illumination from a 10-W light bulb. A disc could store up to 15,000 frames of analogue video, up to 12 h of compressed audio (100:1) or a corresponding amount of digital data. The multitrack read-out facility enabled 30 s of audio to be provided for a frozen frame of video, a capability not available on other current playback systems.

The system appeared to have excellent prospects and potential in a number of applications, especially since the time required to validate an interactive program was considerably shorter than with other processes, due to onsite mastering, a replication capability and rapid turn-round.

However, Atlantic Richfield later decided to pull out of this field of development and put the whole outfit up for sale. The major reason for this decision appeared to be the escalating development costs caused by internal disagreements on policy. Eventually in June 1981 the McDonnell Douglas Corporation took over the operation, including the laboratories and a number of the senior staff. Their initial interest is apparently in the potential of the system for interactive training and flight simulation within their own organization.

Operational players were seen in action in the laboratory and major advances included the possibility of increasing the limit of video frames to 40,000 per side and the use of multilevel pulse amplitude modulation techniques. The 30 s audio provision is achieved by parallel read-out of 10 tracks, one for the video frame and nine for the compressed audio. Copy discs can be either silver-based or diazo. The production target for players is 50,000 per year by 1983 with an estimated cost of $2-3,000 per player.

The initial proposal is for recorders to be provided as a central facility at a cost of the order of $8-10,000, including playback.

4. TERMINAL DATA CORPORATION, WOODLAND HILLS, CA

A further visit to this company was made since it is known that it has developed a series of products which make an impact on the future of information dissemination technology. These products include high-speed document digitizers and high-resolution display terminals.

4.1 The problem with consumer television resolution

All consumer television monitors are optimized to display standard broadcast (NTSC) colour video images. This results in a matrix of dots on the television screen of about 450 pixels (dots) by 450 pixels. When an ordinary document, such as a patent drawing, is shown on a television screen the resolution is too coarse for fine detail and small print to be read. If the details cannot be read from the monitor, they cannot be read from a paper copy printed from the same image.

There is a growing demand for increased resolution. Terminal Data has developed a line of high-resolution VideoMate terminals (Fig.1) which display digitized documents at 200 dots per inch in both the horizontal and vertical direction. (This compares with less than 50 dots per inch on a similar sized consumer television screen.) These terminals interface directly to 200 dot per inch printers. With both the VideoMate terminal and paper copy from the printers, small details such as 4 point type can be easily read. 200 dots per inch is considered the resolution threshold of acceptability for viewing detailed images on a video screen. NTSC video is only acceptable for documents with coarse, large lettering. An added feature is that the VideoMate displays are vertically oriented and can display an 8½-in x 11-in document actual size.

Terminal Data recommends the use of high-resolution terminals for the display of documents stored on microforms, magnetic media, or optical discs.

VMT-2000 VIDEOMATE

KEY FEATURES

- *Full Frame RAM Memory*
- *Wide Range of Data Rate Inputs*
- *High Data Rate Output for CRT Refresh*
- *2200 Line CRT Display*
- *30 Frame/Sec 2:1 Interlace Flicker Free Display*
- *Full Alphanumeric Keyboard to Communicate with Host CPU*
- *1920 Character Capacity (24 Lines/ 80 Characters)*
- *Compatible with Telecommunication Standards*

OPTIONS

- *Image Editing*
- *Gray Scale*
- *8160 Character Capacity (85 Lines/ 96 Characters)*

DESCRIPTION

The VMT-2000 Terminal is a computer data display and a high resolution video raster scan display in the same unit.

The VideoMate terminal displays documents, graphics, computer generated alphanumeric and word processing images. The data display is provided by a companion keyboard and character generator.

The video frame memory and video terminal provides remote raster scan image display. The VMT-2000 includes a 3.8 million bit digital memory which can be loaded from a standard data line. The image memory is then clocked out at a 140 megabit rate for a high resolution, flicker-free image. The 2200 line display is approximately 8½″ x 11″ with a display resolution of 200 dots per inch in each axis.

Terminal Data Corporation, 21221 Oxnard Street, Woodland Hills, California 91367 • (213) 887-4900 TWX 910 494-1918

Fig.1   Model VMT-2000: a video frame memory and high-resolution terminal (Terminal Data Corporation)

SPECIFICATIONS

Refresh RAM Memory

— 3.8 million bit storage, 1728 pixels x 2200 lines
— Video scanner, disc drive or data line input
— Selectable input bit rate 140 K bits/sec to 8 M bits/sec.
— Input format is line sequential bit serial pixel data
— Data rate output is 140 M bits/sec.
— Gray scale (optional)
— Reduced memory (optional)
— Lower input and output rates (optional)

Memory Controller

— Internal clock, crystal or AFC line lock reference
— Format conversion to 2:1 interlace 30 frames/sec. output
— Provides frame and line sync output

Keyboard & Character Generator

— Teletype compatible
— RS 232C or 20 MA current loop interface (others optional)
— Selectable baud rate (110 to 9600 bps)
— Character generator 8 x 12 dot matrix
— Refresh rate 60 hertz
— Full alphanumeric keyboard with separate 10 key pad plus special function keys
— 1920 alphanumeric characters (24 line/80 characters)
— 8160 alphanumeric characters (85 line/96 characters) (optional)
— Image editing (optional)

Mechanical & Electrical

— Overall Dimensions
  Height: 19" (48.3 cm)
  Width: 19" (48.3 cm)
  Depth: 18" (45.7 cm)
— Weight: 36 lbs. (16.3 KG)
— Voltage: 105-120V, 60 Hz
— Power: 190 Watts

CRT Display

— Size 15" (8½" x 11" format)
— Horizontal deflection 68 Khz
— Dynamic focus
— Spot size 0.005" (0.13 mm)
— Video band width response 120 Mhz
— 20° tilt adj.
— 200 dots/inch on each axis
— Phosphor P4 (others optional)
— Gray scale (optional)

CRT Controls

Internal:	External:
Height	Brightness
Vertical linearity	On/Off
Vertical size	
Centering	
Width	
Focus	

SYSTEM DESIGN

The complete VMT-2000 frame memory terminal includes:
— 3.8 million bit RAM video memory
— RAM controller
— High resolution 15" CRT
— Terminal keyboard
— Character generator

The RAM provides a full image frame of digital memory. Information can be loaded into the memory at a wide range of data rates. Stored data is reformatted. It is then clocked out at a 140 megabits data rate to provide a refresh signal for the high resolution display.

The RAM controller directs the loading and unloading of output data. Timing and synchronization signals are included.

The CRT provides a 2200 line image and is scanned 60 times per second in a 2:1 raster scan interlaced format. The image quality is provided by large memory, high refresh rate and high intensity P4 CRT, resulting in a white flicker-free image.

A full alphanumeric keyboard with a numeric 10 key pad and special function keys is available to communicate with host computers. The optional character generator supplies high character density input to the CRT.

VMT-2000 BLOCK DIAGRAM

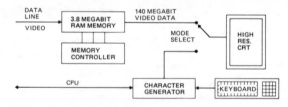

Terminal Data Corporation, 21221 Oxnard Street, Woodland Hills, California 91367 • (213) 887-4900 TWX 910 494-1918

Fig.1 (continued)

4.2 An optical disc production system using the DocuScan 2000

Terminal Data proposed the following system for production of optical video discs containing high-resolution images from original or existing source material.

The Terminal Data DocuScan digitizing system (Fig.2) is completely portable, permitting it to be set up in a central facility or in a customer's facility. Other components can be mounted in enclosures on casters for easy movement. The system can run at a production rate of up to 3200 images per hour. Two operators are recommended to do presorting, indexing and other tasks to facilitate a constant high throughput.

Documents are prepared in order to be scanned. Options allow writing of optical recognition codes or other methods to trigger the start of each group of documents. The DocuScan also accepts command cards for auto control functions.

Documents are fed into the DocuScan one at a time. The operator can select one- or two-side scanning and other functions with a convenient control panel. As each document is scanned, at 200 dots per inch, serial data flow to a buffer plane of the image data compressor. Dual compressors output the digital image data at approximately a 12:1 compression ratio. The compressed image data are transferred direct to a data encoder/formatter. At the same time an electronic pulse is sent to the DocuScan solid state CCD camera, enabling it to start scanning the next image.

The data sent from the compressor to the data encoder/formatter pass through several processing steps and are finally recorded on to the digital optical disc.

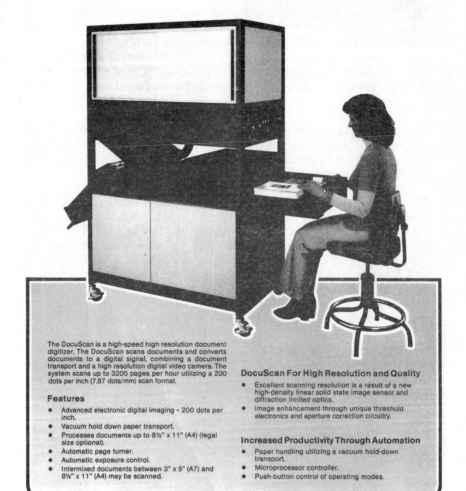
Fig.2   DocuScan: a high-speed document digitizer (Terminal Data Corporation)

Document Handling

A key feature of the DocuScan is the document transport utilizing vacuum hold-down. The operator places a document to be scanned against the alignment edge at the feed station, and the DocuScan does the rest.

Each document is transported to the scan area from the feed station at a speed of 100 inches per second. The document is scanned and passed to the output receptacle, or reversed and transported through a unique page turner. After being reversed, the document is passed again to the scan area for capture of the reverse side. The document may also be turned without scanning the reverse side to maintain original collation.

Automatic operation and the 100 inch per second (2.5 M/S) transport speed combine to provide a throughput of over 3,000 images per hour.

Auto-Command Functions (Optional)

Command cards intermix in the document stack as batch or file separators signal transition from one batch to the next. Command cards are detected but not scanned. This batch or file separator signal is provided over the interface. The command cards are ⅝" (16mm) longer than standard documents providing easy recognition and removal from the output document stack.

Digital Video Camera

The all-solid-state scanning system uses a high frequency 1728 linear diode array sensor. The camera has a horizontal and vertical scan format of 200 dots per inch. The system operates with a 1728 by 2200 line raster to scan a complete 8½" x 11" (A4) document in approximately half a second. The output data rate is 9.24 megabits per second.

Image Quality

An advanced digital video camera, plus its associated optics and illumination system, insures resolution consistent with the highest industry standards. The scanning standard is 200 dots per inch which is compatible with high resolution facsimile systems. Automatic exposure control, a wide dynamic range sensor, and constant illumination insure digitized accuracy regardless of background reflectivity and color.

Specifications

Document Size:
3" x 5"   (76mm x 127mm) — minimum
11" x 8½" (280mm x 216mm) — maximum (14" x 8½" optional)

Document Weight:
0.002 inch tissue   — minimum
0.1   inch card stock — maximum

Throughput:
1.5 seconds per document single side — straight through
2.0 seconds per document single side — turned over
2.2 seconds per document both sides — turned over

Image Quality:
Scan Format — 200 dots per inch (7.87 dots/mm)
both horizontal and vertical
1728 x 2200 pixels (line raster)

Automatic Exposure:
Range — 40% to 100% document reflectivity

Interface:
Drivers and receivers must conform to EIA RS-422A standards using two-wire balanced input. Maximum cable length is 50 feet (762 m).

Output Signals:
End of Line	Ready
End of Page	Busy
Data Enable	Alarm
Video Data	Command Card (optional)
Video Clock (rate of 9.24 MHz)	

Input Signals:
Start Command

Environmental Requirements:
Temperature Range — 60° to 95°F (15° to 35°C)
Humidity   — 20% to 95% relative
Power Requirements — 230 Volt AC, 24 Ampere, single phase
60 Hz (50 Hz optional)

Physical Characteristics:
Cabinet — free standing console
Dimensions — 74 inches (1.88m) high, 29 inches (0.74m) wide
70 inches (1.78m) long without removable
output tray
86 inches (2.18m) long with tray
Shipping Weight — approximately 1000 pounds (450 kgs)

Regional Sales Offices

Eastern .... P.O. Box 234
Fair Lawn, NJ 07410
(201) 444-1773

Washington, D.C. .... 5503 Cherokee Avenue
Alexandria, VA 22312
(703) 750-3717

Mid-Western .... 1100 South Main Street
Lombard, IL 60148
(312) 620-0257

Southern................. 12700 Hillcrest Road
Suite 104
Dallas, TX 75230
(214) 386-6276

Western.................. 21221 Oxnard Street
Woodland Hills, CA 91367
(213) 887-4900

Sold and serviced through a world-wide dealer-distributor network.

Terminal Data Corporation
21221 Oxnard Street
Woodland Hills, CA 91367
(213) 887-4900   TWX: 910-494-1918

Fig.2   (continued)

5. NEBRASKA VIDEO DISC DESIGN/PRODUCTION GROUP, LINCOLN, NE

This group is a specially created unit of designers, producers, technicians and video support personnel, led by Rod Daynes, formerly of Hughes Aircraft. They are dedicated to the development and production of innovative programmes that demonstrate all of the new video disc technology's unique instructional capabilities[4]. The following description of the group's activities has been approved by Rod Daynes.

The Nebraska Video Disc Group was formed in 1978, under a multiple-year grant to the University of Nebraska Television Unit from the Corporation for Public Broadcasting, to investigate and develop video disc technology on behalf of American instructional and public broadcasting.

This group, the only one of its kind in the USA, is charged with designing and producing video disc programming for a wide range of educational and training applications and with developing production procedures for premastering (ie production and assembly of video disc material). The group has the added responsibility of disseminating research on video disc formats, on design and production procedures and on potential future educational and industrial markets for the video disc.

The Video Disc Group is located at the Nebraska Educational Telecommunications Center, headquarters of the nine-station Nebraska Educational Television Network. The Center is nationally recognized for its professional standards and well-equipped studios and production facilities. Through the facilities of the Center it is possible for the Video Disc Group to design, produce, duplicate, store and distribute every form of instructional technology. Production support also includes art, graphic, photographic, film, videography and print services.

The group is developing video discs in accordance with the capabilities of three categories of existing interactive video disc players. Because all video disc players do not have the same features, they

16

are not all equal in their potential for interactivity. Also, a video disc player can be part of a larger system incorporating a computer, further expanding its interactive potential.

We can identify three basic levels of interactive capability for video disc players and player systems:

Level 1 — a video disc player with little or no processing power. This includes the Magnavision models 8000 and VB-8005 CHO1, the Pioneer VP-1000 and DVA TR-7810. These players have no programmable memory, but they are capable of manual interaction with the user. Chapterization, automatic picture stops, frame search and the normal optical video disc features (fast and slow motion, freeze frame, dual channel audio and so on) are available at this level.

Level 2 — an educational/industrial player with built-in programmable memory. Players of this level include the DiscoVision Associates PR 7820 and PR 7820-2 and the Sony LDP-1000. Branching, based on user input, makes it possible to achieve a higher level of interactivity.

Level 3 — a video disc player interfaced with an auxiliary computer. At this level the video disc system functions primarily as a peripheral device, while the computer provides most of the interaction for its user. Very complex branching strategies, and score-keeping, are possible.

5.1 Design and scripting

The process of design and scripting for video discs is considerably different from that for other media. There are several important considerations that must be taken into account when initially approaching the design of a video disc.

First, the type of video disc player that will be used must be determined. As mentioned above the capabilities of the various players are substantially different. A consumer player will simply not do what an educational/industrial player will. Even the educational/industrial players have different features and protocols. Once the player has been identified, the process of media selection begins. Since the disc is an omnibus medium that will carry other media, we can choose from video, slides, film or electronically generated text and graphics. We also have the choice of motion or stills. We often use a single still frame when motion is not necessary, to save running time on a disc.

The chosen media for a particular segment may be determined by such factors as cost and the necessity for perfect freeze frame or slow motion. For example, it is considerably cheaper to generate large numbers of text frames on an electronic character generator than it would be to develop art cards. In the case of freeze frames and slow motion, the only way to get a rock-steady image from the disc is to use film that has been specially encoded or shot at the proper frame scan rate. A necessary step in the design of an interactive video disc is the development of a comprehensive flow chart. A given disc may contain dozens or even hundreds of segments that may be accessed in various combinations. Only a flow chart can reflect the complex interaction between the segments. A flow chart is used at many stages of production from scripting to final assembly. It may be necessary for example to use the flow chart when working with a scene designer or cinematographer to explain the continuity between segments. In short, the flow chart, not the script, is the master blueprint for our interactive video disc.

5.2 Premastering

Before a video disc master can be made, all material that will make up the programme must be put on to a single film or video tape. This is because video disc masters are recorded in real time at 30 frames per second (in the NTSC system), and must contain all this information in a way that is synchronized frame by frame with the disc recorder so that legible disc copies can be reproduced. Since this material may include art cards, slides, photographs, film, video tape and visuals created by an electronic character generator, this process can be rather complex.

Either video tape or motion picture film can be used for premastering and the video disc manufacturers first recommended that, if still frames were to be incorporated on the disc, then the premastering format should be film. It was conjectured that the time difference of $\frac{1}{60}$ s per video tape field might cause an irreversible screen flicker on the resultant disc. The use of film, which is capable of displaying one complete frame every two fields, would thereby eliminate any possibility of flicker. But film is an expensive medium, especially so if the user must convert existing material in other formats to film. It is for this reason that the Nebraska Video Disc Design/Production Group has chosen to premaster all the video disc productions on video tape. They therefore incorporate all of the various media into single tape that can be made directly into a video disc master. Such a complex, video-based premastering system requires a sophisticated, computer-controlled editing system. The Nebraska Group is in a fortunate position in that the educational television network has such a system.

5.3 Video disc simulators

Since the video disc is a read-only medium, developers of disc material have two alternatives to test the effectiveness of this material. Either a limited quantity of discs must be mastered and replicated and then tried out on a player, or a simulator must be constructed that tests the material prior to the commitment to the mastering and replication process. The first choice is expensive and not very practical. Disc mastering and replication, even in limited quantities, is not cheap, but more important, the turn-round time can vary from two weeks to as long as six months. This can cause logistical problems with various scheduling and studio times. If mistakes are found, then remastering will be necessary, which doubles the cost. On the other hand, if a simulator is used initially as a formative evaluation device, then the logistical problems and cost can be reduced considerably. The Nebraska Design Group has developed a video disc simulator for such purposes. It has taken a three-level approach to evaluation at the various stages of the design, scripting and production process.

The first level is a paper-based system using story boards and flow charts. At this level it can check sequencing, branching strategies and overall design.

The second level involves the use of a U-matic video cassette recorder controlled by a microcomputer. The programme is transferred to cassette and the computer controls the playback in the same sequence as that on the video disc.

Finally, it has developed a third level simulator using a Bosch BCN50 1-in VTR under computer control. This system gives a very rapid access (for a video-tape-based system), some slow motion capabilities, excellent freeze frame through the VTR's still store, and single-frame-still step forward and reverse. In addition, since the Bosch machine is a high quality VTR with a built-in editor, stills or motion sequences can be re-inserted or changes can be made to the wording of a text frame.

The group essentially offers a premastering service, ie including all steps up to, but not including, actual video disc mastering. The group does not manufacture video discs; that service is provided by a mastering/replication facility such as those operated by Disco-Vision Associates, Sony or 3M.

The group does not like the DiscoVision process of producing two steps of video tape from the film stage, or one more even if sent a properly formatted tape. It causes a faster degradation of quality which it feels is already bad enough due to the poor quality control in disc mastering. It prefers the 3M photopolymerization process, which is now on stream and claimed to be providing better quality discs. Although the group has done little work on digital coding, it considers that the 3M process will be able to cope.

It has produced a computer interface to the Apple computer which allows overlay of computer information and graphics on a video terminal rather than the usual practice of alternate format. This interface will be commercially available within a year for $500-700.

The author discussed with the group the possibility of preparing a tape from the specimen COM film of the catalogue which it inspected. It would be prepared to do this, but at a cost of 50c per frame. This charge seems high, even though a 1-in video tape machine with single frame editing costs about $150,000.

6. MASSACHUSETTS INSTITUTE OF TECHNOLOGY, CAMBRIDGE, MASS

Visits were made to two renowned laboratories at MIT as reported below.

6.1 Laboratory for Information and Decision Systems

This laboratory is under the direction of Professor J. Francis Reintjes who was the leader of the Project INTREX group in the early 1970s when much pioneering work was done on automated fiche retrieval systems. Professor Reintjes was at the Cranfield conference in July 1981 and invited the author to visit the laboratory to see its work on document digitization and transmission for library applications.

The position taken by Reintjes is that, in order to take full advantage of new digital technologies, one should think and plan in terms of entirely new procedures for handling library-type information, rather than merely moulding the new technologies piecemeal into traditional operations.

One example of a new approach is interlibrary resource-sharing among a group of libraries interconnected electronically by means of a digital-communications network. In this configuration, the content of documents would be moved electronically between network nodes, rather than the physical documents themselves or photocopies of segments of documents. Resource-sharing through electronic networks most likely will be applicable to serials holdings rather than to books, since interest in serials frequently pertains only to an individual article which can be quickly scanned and transmitted.

Interlibrary resource-sharing networks for serials holdings are attractive because packing density on discs is high if information is stored in digital form. The speed of response to interlibrary requests is also fast; 'within moments' for a complete transaction of information already in digital storage is achievable and 'within the hour' is possible for transactions of printed material that must be scanned and digitized on demand.

Clearly, in the light of such quick response times, it is appropriate to question the need to replicate all serials holdings at all nodes of the network. One should also begin thinking about ways to use the interlibrary network transactions records as a means for determining the most effective locations for serials holdings, and about how the data-gathering feature of the system can be utilized for weeding-out purposes or to relegate certain materials to a second- or third-level demand status.

How information might flow between two nodes of an electronic resource-sharing network is indicated as follows. It is assumed here, for library-control purposes, that all requests for serials flow through the system. In order to obtain a document, the user fills in a request form. (An alternative approach is to have the user insert the request himself or herself at a computer. This approach might be introduced into the system later, after the experience has been gained with network procedures.) The form is given to a library clerk who verifies its correctness. If incorrect, it is returned for correction; otherwise, the information is used to determine whether or not the user's own library owns the document. This step is a computer operation, since a complete index of all serials held by network members will be stored in a computer. Should the document be owned locally, the form is given to the user who seeks it in the stacks. The process ends if the document is found; if not, the user resubmits the request form. It is also possible that the local library does not own the document. In either negative case, the clerk ascertains from computer-stored information which, if any, libraries in the network hold the document in question; if none does, the process ends. If one or more do, the clerk enters request data into the computer. The amount and kinds of data that must be posted remain to be decided; in general, the data will depend upon how automatic succeeding steps are. Two limits are envisioned. At one extreme, the clerk transmits a request message via computer to the library most likely to have the item being sought, repeating the step, if necessary, for the next most likely source and so on until a supplier has been found. At the other extreme, with a more elaborate computer file structure, further clerical action beyond the initial entry of request would be unnecessary except to check on the status of request. The computer would automatically transmit the request serially to various libraries until the request was honoured.

If no supplier in the network can be found, the process ends. If a

22

supplier is located, the request form is delivered to the local transmission operator by the requesting clerk for identification purposes, at which point the local clerk's task ends (except perhaps to enter billing information into the computer on receipt of notice that the document has been sent to the user).

Action is now transferred to the supplier's location. The supplier's clerk receives the document request from the resource-sharing computer and determines whether or not the item is on the shelf. (We know the supplier owns the document; otherwise, that particular supplier would not have received the request.) If the answer is NO, that message is inserted into the computer and the next most likely source is tried. If the item is available, it is de-shelved and given to the transmission operator together with the request form. The transmission operator in turn electronically scans and transmits the document, after which it is returned to the clerk for replacement in the stacks.

Action now reverts to the requester's location. The transmission operator there receives the transmitted document, informs the user by phone of its availability, or delivers it by internal mail, and finally returns the original request form to the clerk, who posts accounting data into the computer.

The process as outlined above is for illustrative purposes only. As implementation problems are uncovered through modelling of computer hardware and software configurations, alternative information flow model processes will be examined.

the laboratory a bound-document scanner has been developed. A unique feature of the scanner is a flexible "librarian-friendly" book cradle which allows the document being scanned to be put in position with minimum stress on the binding. A Fairchild 2048 pixel CCD array has been mounted on a travelling frame which moves across the horizontal plane of the bound-document page giving 1728 scans in about 5 s. No data compression was applied but they talked of using a run length coding scheme expecting to get an average compression of about 4:1. At the receiving end of the system the signal was fed to a Gould electrostatic printer with an 11-in (approximately A4) span giving a corresponding horizontal traverse of the document. The development is encouraging in that a resolution capability down to 6-point type is being achieved.

The other major activity in the laboratory is an investigation of a translating computer interface as a means to simplify access to, and operation of, heterogenous bibliographical retrieval systems and databases. This is a comprehensive terminal intended to link 'naive' users into any database and the interface allows users to make requests in a common language, the request then being translated into the appropriate commands for whatever system is being interrogated. System responses may also be transformed by the interface into a common form before being given to the users. Thus the network of different systems is made to look like a single 'virtual' system to the user. An experimental system named CONIT was built and tested under controlled conditions with inexperienced end users. A detailed analysis of the experimental usages showed that users were able to master interface operation sufficiently well to find relevant document references. Success was attributed in part to a simple command language, adequate online instruction and a simplified natural-language, keyword/stem approach to searching. It is concluded that operational interfaces of the type studied can provide for increased usability of existing systems in a cost-effective manner, especially for inexperienced end users who cannot easily avail themselves of expert intermediary searchers. Furthermore, more advanced interfaces based on improved instruction and automated search strategy techniques could further enhance retrieval effectiveness for a wide class of users. These issues are currently being addressed by the research group.

6.2 The Architecture Machine Group

This group, led by Professor Nicholas Negroponte, is investigating techniques for improving man-machine interaction. Its work is of an exploratory nature rather than more traditional hypothesis-testing style. A central facility in its laboratory is the Media Room, used in the development of a 'spatial data management system'.

Essentially the Media Room is a luxuriously furnished room approx 20ft x 10ft, one wall of which has been completely replaced by a back projection colour TV system. In the centre of the room is a comfortable armchair with 3-in square pressure sensitive tablets mounted in the arms. To the left and right of the chair within arm's reach are colour TV monitors with transparent touch-sensitive panels fixed across the screens. In the corners of the room are loud speakers providing octophonic sound. Speech input is achieved via a microphone mounted in a headset connected to an off-shelf speech analyser.

Perhaps the most novel part of the system is the 3-D position-sensing hardware. The user has a small box, the size of a sugar cube, strapped to his wrist or some other part of his anatomy. This contains apparatus which senses three orthogonal rotating magnetic fields, the position and orientation of which are established by another, larger box which stands on the floor. This particular piece of hardware can be used in much the same way as a 'mouse', to drive a cursor across a large screen or TV monitors, except that in this case the user controls the movement by waving his arm about or pointing.

The user model of the spatial management system consists of an apparently infinite space in which various different data objects are located. The user is able to 'move and zoom' into the object using either the touch parts in the chair arms, voice commands or waves of the hand. Objects in the system include books, a TV set, a calculator, maps and a video disc player. Some objects make noises and as one 'zooms' towards them the noise becomes louder. The user can interact with objects in a variety of ways. For example an object may display a menu on one of the small TV monitors (these monitors augment and clarify the image displayed on the large screen) and the user can pick something from the menu by pointing to it on the screen. Alternatively he can speak to the object.

In a graphics laboratory such as this the value of the industrial video disc player as computer-controlled memory system was quickly realized. A fairly sophisticated application was implemented in connection with the Mapping By Yourself project[5]. The idea was to allow the user to get to know his way around a city. A research team went to Aspen, Colorado, with cameras mounted on top of a van. They drove up and down all of the streets in town, filming one frame every 10ft. Then all turns at all intersections were similarly filmed. Next 35-mm slides were shot of all buildings and some included an in-depth slide tour. For example, pictures might be taken of a restaurant's dining room and menu as well as its head chef and speciality dessert. All film was edited and transferred to video disc. The resulting disc is a database with which the computer can take the user round town under his control.

Briefly, the Travel program works as follows. Two video disc players are used so that one is being viewed while the other one is anticipating the next sequence and seeking to it; this is known as staging. For example, you are travelling east on Main Street. Upon reaching an intersection, the computer notes that you have not taken any action and so continues to play through the intersection. The user then indicates that he would like to turn right at the next intersection. Under program control, the other disc player searches for the starting frame of the desired upcoming turn. This frame number comes from a database contained in the computer. When the intersection is reached, display is cut to the other player and you make a right turn on to Monarch Street, after which the 'south on Monarch Street' sequence is shown. The user also has the option of varying speed and direction (forward or reverse) and of stopping. He may also stop and touch a building on the screen. A touch-sensitive display indicates to the computer where the user touched and that information plus a digital database is consulted and the front view of the desired building is searched to and displayed. If that building has a slide show associated with it, the user can ask to see it. Thus the combination of the video disc and the computer allows dynamic insertion of building façades and user-controlled routes.

To familiarize you with Aspen, you could watch a video tape of someone else's ride through town, stopping at places he thought you might like to see, but the Travel program provides for a large, general and flexible database that you can use to fit your needs and interests.

In order to organize, access and control 54,000 frames, there must be some sort of description of what is where. For example, in Travel the computer has to know what the starting and ending frame numbers are for the right turn from Main to Monarch. Given that there are visual data for each turn and street and every building in Aspen, the digital database that must accompany that disc becomes quite large, since the computer must know where things are within each frame.

Although the databases are large, if a particular frame is being displayed, then only a particular portion of the database is useful at that time. For example, the frame number of slide shows for the buildings in Aspen are needed only for the subset of all buildings currently in view. By replacing the picture information at the top and bottom edge of the screen with encoded digital information, the video and digital databases combine into one medium. This database expands to include full frames of encoded text for computer display, program code or any other useful data. Thus the goal here is to provide a low cost, NTSC-compatible system for storing both visual and numeric data.

In order to determine effective encoding schemes for impressing the digital data on the video disc the group has studied the various distortions that limit data rates and accuracies. By far the most frequent cause of error on discs is dropouts. Dropouts observed here range in size from microscopic to as large as 0.15 mm across. Unlike additive Gaussian noise, dropouts are impulse noise sources that are independent of the data modulation technique used and effectively yield a signal-to-noise ratio of zero.

A prototype model of a disc-to-disc decoder was modified to accumulate statistics about dropouts. In order to carry out an analysis over many frames the player must be under computer control. The DiscoVision PR 7820 Model 2 supporting a Universal External Interface was used, which allowed an RS232 port from the computer to issue commands and to request status in order to monitor the system. Test programs were developed to search to a start frame and accumulate dropout statistics for some number of frames. From this information worst case error patterns were chosen and appropriate error control schemes developed. Of a relatively small sample of 1,600 frames the worst frame exhibited about 270 dropped-out bits. Assuming 480 lines x 400 bits, ie 192,000 bits per frame, this results in one error in 711, worst case. The longest single dropout was 96 bits long.

Using the dropout statistics as an approximation of error behaviour, an error-correcting code can be selected. Other criteria include encoding cost, decoding cost, expected unrecoverable error rate, data density (bit clock rate) and decoder speed. There are many trade-offs to be examined and most are dependent on the application. Since one goal is to provide data supplemental to visual images as they are viewed, the encoding block size should be small, no more than 5 percent of the screen or 24 lines. Since 'proof reading' discs after production is impractical, it is important to keep the error rate down. Even after an unrecoverable error is detected on a disc, it cannot be patched and must be discarded. Assuming the application calls for 24 lines per frame, a reject rate of one error in 100 discs might be reasonable. Owing to the publishing nature of the disc, ie the discs are decoded many more times by many more people than they are encoded, encoding cost becomes less important. Lowering decoding cost is attractive because it would make large information bases available to relatively small, inexpensive computing systems.

The group has done a considerable amount of work in the selection of a suitable error control and modulation system and has built a prototype digital data decoder board for evaluation of the many problems yet to be solved. (Performance characteristics of 200 user bits per line, corrected to one error in 10^{12} bits, are currently being talked about.) The group believes that research on digital data on video discs is not being pursued very actively by the video disc manufacturers or the data processing industry. What little is being done is by potential end users.

7. NATIONAL LIBRARY OF MEDICINE, WASHINGTON, DC

Owing to the airline traffic controllers' strike it was not possible to make the intended visit to the NLM. However, information has been received from the group at the Lister Hill Centre regarding the progress of the project involving the design of a prototype system that will electronically scan, store, retrieve and display the full page of each document required by the library. It is understood that the long-term goal of the programme is to introduce advanced technology to help the library fulfil its mission as a national archive for biomedical literature, as well as its responsibility to maintain an interlibrary loan service.

Currently the loan service is handled manually — original volumes and photocopies are sent by mail. This process is proving inadequate because of the large and growing volume of loan requests. The NLM also foresees a number of problems if the library continues to store all its materials in print form. In 5 to 10 years, it just won't be possible to do business as today — storing paper, photocopying and mailing.

Despite the developmental and evaluative aspects of the project, once the system is operational and assuming that there is a national broad-band communications network at that time, other libraries will be able to access the full text, including illustrations, of biomedical documents stored at the library.

In addition to housing print and film materials, the library offers an online system of biomedical databases, called MEDLARS. The system contains citations and abstracts from literature on toxicology, population, health planning and administration, and biomedicine. There is also a database that has information on about 2,000 substances that are of known or potential toxicity, and to which substantial portions of the population are exposed.

The techniques being explored in the new project might eventually be used for the MEDLARS system. The NLM is already working on development of what it calls MEDLARS III in which it hopes to automate many of the activities surrounding acquisition of database materials, cataloguing, indexing, and so on.

At some later date, perhaps by the time of MEDLARS IV, the NLM might decide that the medical community should be able to get the full text, but it is too early to say what impact this project will have on the MEDLARS system.

7.1 Electronic document storage and retrieval

The experimental system is to be developed by pursuing three concurrent research projects — document capture, data transfer and storage, and document display.

The document-capture project will develop techniques to scan electronically paper documents and film media containing textual and graphic material, and to digitize the analogue electrical signals generated by the scanning process. Paper documents include both bound volumes and separate sheets of paper. The system will also be capable of accepting machine-readable data.

In medical literature, from 30 to 50 percent of the journal pages have at least some graphic illustration on them and they are critical to understanding the whole text. This is the reason that medical literature in full text requires scanning the facsimile of the page rather than just keying in the textual information, as is done with other full-text systems now in operation.

The scanner of the system is to be made up of an array of 2,048 charge-coupled devices (CCDs). These are photo-sensitive detectors that will convert the black-and-white image of a page into electrical signals. With an array this size, the "scan density" on a page is about 200 dots per inch.

The NLM is not sure that 200 dots per inch will give the resolution needed to allow proper legibility, but thinks it is the minimum for the type of printed material with which it has to work. This is one of the questions the NLM hopes to answer by building the prototype system.

The use of scanning in the system also explains the need for a broadband communication network if outside users are going to access library materials online. A single page with approximately 3,000 characters of information can be represented in the American Standard Code for Information Interchange (ASCII) by about 24,000 bits, but represented in black and white elements (exact facsimile), the same page would require 3.74 million bits. This can take a considerable amount of time to transmit over a narrow-band channel.

However, once a broad-band network is available — and several companies are working toward this — with a channel capacity in the megabits region, it will become feasible in terms of cost and time to transmit scanned pages.

The data transfer and storage project will develop techniques to control the flow of the digital signals produced by the scanning and digitizing processes as these signals are transferred to magnetic discs for temporary storage. The system will be able to retrieve data from the magnetic disc storage for display, as well as for processing prior to archival storage on high-density optical discs.

Compression is an example of such processing. It will help minimize the storage and transmission problems. The scanning and digitizing processes deliver signals containing a considerable amount of redundancy. Subjecting these signals to suitable forms of compression results in fewer bits that represent essentially the same amount of information.

It is estimated that by using a compression ratio of 20 to 1, more than 500,000 journal pages could be stored on a single optical disc. A "jukebox" or disc pack containing 1,000 discs could provide an online storage capacity of three to four million journal issues.

The document display project will develop methods to display retrieved documents in both softcopy (electronic) and hardcopy (paper) forms.

Although the initial three R & D projects will develop and integrate the components of an experimental system for evaluation within the NLM, a follow-on phase is proposed to investigate disseminating the stored documents nationally.

7.2 Video disc program

The Lister Hill Centre's video disc program includes several inter-related activities. The Communications Engineering Branch is developing a Video Processing Laboratory to provide the facilities necessary to create master tapes for video disc production. The Centre's Health Professions Applications Branch and the National Medical Audiovisual Centre are concerned with the effective use of this exciting new modality. The Computer Technology Branch's responsibility is to develop an intelligent Video Disc Interface Unit, and to identify and address the problems associated with encoding digital information on the standard, video-formatted video disc.

The viewer of television images from video discs is relatively tolerant of coding errors; encoded digital information, however, will have to be essentially error free. Studying the encoding of digital information on a video disc is primarily a problem of determining the errors introduced during the mastering and replication processes and, subsequently, resolving these problems. The initial emphasis has been on developing, with the cooperation of industry, an experimental facility to determine the error characteristics of present video disc production.

The first objective is to devise an error-correction coding scheme that can be implemented on an inexpensive microprocessor and incorporated into a single "black box" with the Video Disc Interface Unit. This is a necessary step toward the long-range goal of using video discs for the mass publication of large machine-readable databases, as well as compendiums containing digitally encoded text, randomly accessible colour video images and audiovisual sequences.

8. SOCIETY OF AMERICAN ARCHIVISTS' 45TH ANNUAL MEETING, BERKELEY, CA

It was intended that a visit should be made to the Public Archives of Canada in Ottawa to discuss progress with Dennis Mole of the video disc pilot studies project in the Public Archives. However, the equipment was being shipped to California for the 45th Annual Meeting of the Society of American Archivists, so arrangements were made to contact Mole there and attend the session on the impact of video disc technology on archiving problems. This session was arranged because the organizers saw the potential of the video disc for supplanting many techniques used by archivists to store and retrieve information. They felt it could already equal the cost of microfilming with the advantages of permanent colour reproduction and high-speed retrieval. Linked with appropriate software, it could assist archivists in rearranging collections to give multiple points of access. It also could provide enhanced details of documents photographed from any perspective and thereby create a three-dimensional record as well. It was intended that two projects should be discussed, with practical demonstrations. They were the Public Archives of Canada project and a project at the Smithsonian Institution on the storage and retrieval of photographs.

It was clear that the Smithsonian project was at a very early stage, mostly a paper study so far, and the demonstration related simply to the capabilities of the DiscoVision PR 7820 player.

Unfortunately Mole's equipment did not arrive from Ottawa in time for the meeting. In a project initiated by the Public Archives of Canada, Mole has been investigating applications of analogue video discs for access to large amounts of archival material using the Thomson CSF transmissive optical disc system. He has developed a system whereby a large number of frames in an image storage system can be interpreted as a collection of orderly arrays of frames through which the user can move with the aid of a minimum number of keys whose function seems intuitively natural. The image storage system holds a large number of images, many of which contain too much detail to be adequately viewed in their entirety on a practical display system such as the standard television monitor. In other words, because of limitations of display technology, it will often be necessary to take a large and detailed image, such as a map, and record and/or display it in sections. In this way the user will be able to resolve the smallest detail on the image, even though the display system cannot present the whole image at the same time as it provides the required resolution.

The result, however, is to multiply further the number of frames the user must explore in order to find the one required. Mole has spent considerable time developing an organizing principle for exploring the contents of an image storage system in which many of the objects are depicted by more than one frame. He divides the larger image into smaller frames arranged as an orderly rectangular array of sub-frames. Some degree of overlap is provided between any one frame and its neighbours. This is done in order to ensure that no area is lost in the process and that users are given some common reference features as they move from frame to frame in viewing the image.

The samples included prints, paintings, drawings, photographs, paper documents, medals, posters, maps and motion pictures. Because there was such a difference in the size of the objects it was finally concluded that the objects should be taken one by one, measuring them in millimetres, then using a printer's glass to measure the finest line for resolution. Some items could be photographed in one shot while for others, like maps, hundreds of shots were needed. These multiple shots were arranged in a matrix form: that is, rows horizontal, columns vertical, so that a three by three matrix would give a total of nine. Each matrix was preceded by one shot of the whole object for orientation. In order to have an appreciation of the size of this task, it should be pointed out that there were 110 maps in the collection which, if a full matrix has been undertaken on each one, will have amounted to something like 16,000 individual frames for the maps alone. The images could not be butted in the matrix because there was a danger of losing something; as a result an exact overlap for each item had to be calculated. Charts were made with television safe image areas so that it would be known exactly where to position the animation XY table and camera.

The photography took three months to complete and altogether some 4,600 frames were taken, appearing as individual frames on the discs. The resulting film strip was spliced on to segments of a motion picture film. The software also had to be written for the computer catalogue and search programme.

Having completed the initial pilot study, Mole and his colleagues have made a further study of the feasibility and practicality of implementing the technology, in particular as an archival recording medium for the Machine Readable Archives (MRA) and the National Film, Television and Sound Archives (NFTSA). Their review of previous studies, the current situation and the capabilities inherent in the DOR process led them to believe that implementation of the DRAW (Direct Read after Write) technology when available will be both viable and desirable. Their reasoning was as follows:

a) archival records would be recorded on a medium which, because of the use of digital encoding techniques, lends itself to unlimited reproduction with no degradation in the quality of its content in successive generations;

b) the capability to produce inexpensive copies of optical discs through the 'one-off' process would enhance the chances of the archived records surviving if a number of copies were distributed throughout the country, to records centres for instance;

c) in as much as optical discs may be stored in normal office environments, in areas with flooring which has a low load-bearing capacity, the requirement for costly environmentally controlled storage areas will be reduced;

d) the amount of storage space required will be reduced in direct proportion to the extent to which the existing copies of records are discarded once acceptable copies of the records are created on optical discs;

e) the copying of MRA and NFTSA holdings on to optical discs will not require significant additional manpower because the process is technology-intensive as opposed to manpower-intensive;

f) the copying of MRA holdings on to optical discs will cause no decrease in the quality of the records being transferred. The copying of NFTSA holdings will cause no greater loss of quality than would be experienced in recopying those holdings on to fresh film stock. The possibility of any loss in quality in this respect is counterbalanced by the potential which exists for image enhancement during the transfer process. Tests conducted to date have indicated that the process is well within acceptable standards for quality;

g) records on the optical disc may be copied back to the medium from which they originated with no reduction in quality;

h) the use of current conservation measures will be reduced in direct proportion to the extent to which existing copies of records are discarded (once acceptable copies of the records are created on optical disc). Those copies of existing records retained in addition to the optical disc records will obviously continue to demand the same degree of conservation as they do at the present time.

Some further elaboration on this last point is warranted however, as adoption of the optical disc as an archival medium implies a total commitment to the conservation of the records retained on this medium. First of all, it is necessary to realize that this medium will have a finite life expectancy. Although accelerated ageing tests and extreme temperature and humidity tests have been performed, estimates varying from 10 to 100 years are still being quoted by various sources for various types of discs. Given normal storage conditions, one can probably expect something between these extremes. The implication of this fact is that it will be necessary to monitor the condition and the integrity of the discs closely on a continuing basis. Secondly, the material on the disc will not be subject to a soft degradation, ie the image will not gradually deteriorate or fade. As this technology uses digital encoding techniques supported by error-correction coding the record will either be there, and reproducible, or it will not. Thus the survival of the record depends on there being multiple copies of each optical disc so that if one copy fails another copy can be used to make a replacement disc for the unusable one.

9. CONCLUSIONS

The impression gained by the author of progress towards digital applications of video discs was disappointing. It is fairly clear that, at least at the present time, there is not much push either from the disc and player producers or from the general data processing industry. The small amount of activity that is taking place is by potential end users. There are several reasons for this situation.

First, with no updating capability and with the necessity for discs to be mastered and replicated at a central facility, the video disc has little application as a conventional computer peripheral, eg as a digital backing store.

Secondly, at the time of the visit, DVA was the only company able to provide players and discs and because of a backlog of production of conventional analogue video discs, due to the necessity to satisfy consumer market requirements and the problems of poor disc yield, it is only just beginning to pay attention to the more specialized potential applications. Since the visit, 3M has announced a disc replication facility using the improved photopolymerization process and with a much shorter claimed turn-round which should help to boost the digital developments. Thomson CSF is also offering a very short turn-round, but the lack of availability of players is still a major problem.

Despite these problems there is considerable scope for developments of the video disc system for digital applications and, in particular, the photographic disc with its more conventional processing requirements has significant potential applications.

10. RECOMMENDATIONS

The recommendations made by the author in the original report to the British Library are still valid. Since there is still no likelihood of the DOR disc system becoming available in the near future, potential applications of the optical video disc in information storage and retrieval should be pursued. The video disc should be looked upon, at least in the first phase, as complementary to microform in automated remote-access storage and retrieval applications and both analogue and digital formats should be included in the study. The programme should also include a study of the potential of both the laser ablative disc and the photographic disc.

11. REFERENCES

1. Barrett, R. *Developments in optical disc technology and the implications for information storage and retrieval.* British Library R&D Report 5623, June 1981.

2. Barrett, R and Sam Ming Hoi, J R. *Remote document storage and retrieval studies.* Report to the British Library (in preparation).

3. Ives, H D. *Type fonts for video presentation.* DiscoVision Associates preliminary report, February 1981.

4. Daynes R et al. *Video disc services of the Nebraska Video Disc Design/Production Group.* University of Nebraska-Lincoln, April 1980.

5. Lipman, A "Movie-maps: an application of the optical video disc to computer graphics". Proc. SIGGRAPH '80, Seventh Annual Conference of Computer Graphics and Interactive Techniques, Seattle, WA, 1980.

12. STOP PRESS

As this report was going to press in May 1982 it was announced that IBM and MCA were pulling out of their joint video disc project with Pioneer in the consumer market, with Pioneer buying their interests. The reason given was the difficulty being experienced in selling video disc systems against the competition from the more flexible video cassette recorder. Pioneer will continue to market players and records in competition with RCA and Philips. IBM said it was pulling out because the market was not growing fast enough, reflected also in RCA's failure to achieve its target for video disc player sales. It will be remembered that, in 1979, IBM and MCA formed DVA with equal shares and this company in turn took a half share in Universal Pioneer, the Japanese company taking the remaining half share with an estimated joint investment of $100 million. DVA used the plant taken over from MCA to manufacture discs (section 2), while importing players into the USA from the Universal Pioneer plant in Japan. The DVA plant in California has since been closed, having produced over 2 million discs, and marketing of players and discs is to cease. DVA remains in existence with a very small staff to handle patents, technology and Universal Pioneer royalties.

Also, 10 years on from the first announcement and after several postponements, the Philips Laservision optical video disc system is to be launched on the UK market at the end of May 1982. In parallel with their commitment to develop the consumer market, Philips says that further substantial resources will be dedicated to entering the interactive application of Laservision for industrial, commercial and education purposes.

It thus seems that the commercial/educational market, which at one time seemed too specialized to support any form of non-recording video disc, now appears the most likely sector to benefit from its fast-frame-access facilities.

LIST OF ABBREVIATIONS

ASCII	American Standard Code for Information Interchange
CCD	Charge-coupled device
CLR/HLT	Clear/halt
COM	Computer output microform
DOR	Digital optical recording
DRAW	Direct Read after Write
DVA	DiscoVision Associates
IBM	International Business Machine
MCA	Music Corporation of America
MEDLARS	Medical Literature Analysis and Retrieval System
MRA	Machine Readable Archives
NFTSA	National Film, Television and Sound Archives
NLM	National Library of Medicine
NTSC	National Television Systems Committee
RCA	Radio Corporation of America
SMPTE	Society of Motion Pictures and Television Engineers
TDC	Terminal Data Corporation
UEI	Universal External Interface
VTR	Video tape recorder

OTHER REPORTS

This is the seventh in a new series, Library and Information Research (LIR) Reports, which are superseding the priced series published by the British Library Research and Development (BLR&D) Department since 1975. The BLR&D Report series will now consist only of unpriced reports deposited at the British Library Lending Division. These reports will continue to be available on loan, or may be purchased as photocopies or microfiche. It will still be possible to purchase reports previously published in the BLR&D Report series from the Lending Division. Details of the first six LIR Reports are given below.

The on-line public library
LIR Report 1   ISBN 0 7123 3002 X
This report describes the establishment, operation and evaluation of the bibliographical information retrieval on-line service provided for the general public by the Lancashire Library on an experimental basis. Detailed information is provided on project management, searches conducted including times and costs, publicity methods, knowledge and reactions of library staff, user characteristics and evaluation of the services, and document availability and the impact on inter-library loans. Supplementary information is included from a sample survey of staff at the Birmingham Public Library.

Prestel in the public library: reactions of the general public to Prestel and its potential for conveying local information
LIR Report 2   ISBN 0 7123 3003 8
An umbrella co-operative, developed by LASER, involving Prestel sets in 36 public libraries and voluntary organizations during the first year of the full-scale public Prestel service is described. The report seeks to simplify the establishment of similar schemes and to educate potential information providers about Prestel.

Searching international databases: a comparative evaluation of their performance in toxicology
LIR Report 3   ISBN 0 7123 3004 6
The seven major databases in toxicology are compared and evaluated according to the criteria of recall, precision, overlap and currency. In conclusion, guidelines are offered indicating which databases are most useful in particular situations.

The use of patent information in industry
LIR Report 4   ISBN 0 7123 3005 4
This report examines, through case studies, the ways in which patent information is used by individuals in industry and in academic research teams. The fruits of an effort to publicize the patent holdings of Newcastle Central Library are discussed and related to the findings of the case studies.

The state of public library services to teenagers in Britain 1981
LIR Report 5   ISBN 0 7123 3006 2
Looking at the reading habits and tastes of British teenagers, this report scrutinizes the response of public libraries to their demands. Policies and attitudes are discussed, and recommendations are made for both current practice and the future.

Prestel in the library context
LIR Report 6   ISBN 0 7123 3009 7
Two seminars were held, in November 1981, on research involving Aslib, LASER and several public libraries. The uses of Prestel as both source and vehicle of information are discussed.